Zen And The Art Of Teaching

A Survey of Techniques With Efficacy

Edward Seymour

Zen And The Art Of Teaching
A Survey of Techniques With Efficacy

Edward Seymour

Acknowledgments

I owe a special gratitude to my sister, Sue as she inspired my passion for math and science as well as learning. In addition, my grandmother held the firm belief that can is a verb, can't does not exist. The list of teachers with whom I have worked would occupy most of the text so suffice it to say, thanks.

Conceptual Understanding

I will open with this disclaimer. I am trained as an engineer who has studied Microelectronics, Photo Science, Economics, Physics and Business. Thus, I did not take classes in "education" but have had significant involvement in programs including National Science Foundation and outreach programs for Math and Science (STEM, National Engineers Week, Society Of Women Engineers and various universities) Thus, the basis of this book is empirical and not theoretical.

It is my belief that to properly teach a topic requires a conceptual understanding of it.

Know The Audience

Given the aforementioned conceptual comprehension, to teach means you need to quantify the target audience and ideally understand "what makes them tick". Essentially a study of the student's culture will help immensely.

Enumerate The Predicates

So that the ease of conveying the concepts and subsequent efficacy is at an optimal level, be explicit about any prerequisites of homework which is elemental to the discussion. A sign of an exceptional teacher is one who has such a grasp of the topic and understanding of students, that little to no prerequisite material is required. This approach allows one to introduce a concept in its most elegant form without preconceived notions or derivations.

Derivations can be left to supporting study when a student questions the basis of the "answer". This approach allows one to focus on the fun stuff.

What are Predicates

How are they defined

How do you minimize them to maximize the students

Define The Steps

This is akin to breaking a concept into digestible chunks or pieces. The elements to be covered here are based on working vocabulary of the student. In some sense, the most elegant answer to this is to pose this in the form of a demonstration experiment where the outcome is tangible and the language is minimized. In this realm I created a very simple experiment to convey the idea of modern physics to a classroom of 2nd grade students. Within a single class of 45 minutes they were able to express an actionable understanding of pn junctions, threshold voltage and even avalanche breakdown. This same experiment was later replicated for use at undergraduate studies.

Enumerate The Steps

Cast learning steps in a rote order for learning (in aforementioned this is problem statement for an experiment)

Assign homework

Define the iteration count you need to have the students so that mastery can be achieved, how?

Iterate On Homework

Spin in the loop on the homework

Check for comprehension

In essence this implies to first "ask the student what they heard" If in vector terms, the answer is way off, adopt a new method of delivery. Typically there is a dominance in visual or verbal reception. This can be big picture or tiny steps.

Assess Mastery

Determine when a student is largely independent. At this point the time has come for a new course of study or an in depth analysis of a gnarly problem. The intention here is to further the state of the art.

Seed/Advise Path

Upon completion of the course, interview the student on aspirations and help set a force vector that defines direction of continued study.

Subjects Taught This Way

Math in English, Signed English, ASL

Over the last 50 years I have tutored, taught or mentored math. An overriding theme has been to find a relevant scenario where the benefit of learning math outweighs the pain of learning. Most students befriend math and even embrace it if there is some glimmer of how as to how this arcane practice relates to life as we know it.

The range of students have included k-12 initially. Then I have included engineering students and recent graduates. The universal observation in this area is regardless of level, there tends to be high value in knowledge of why to learn this. Effectiveness is also greatly enhanced by forming this as an experiment or design challenge. It is far more efficient to enter into a open ended experiment with very few rules. In this way, a natural selection of math techniques results. Thereafter, all one needs to do is spend 5-10 minutes decoding which math skills led to the solution. In this way the student gets really excited to practice the craft again.

As for the reference to Signed English and ASL, these are two forms of sign language "spoken" in the deaf community. I took a position for 2 years time where I was tutoring math at the National Technical Institute for the Deaf (one of the then 9 colleges of RIT) Sign language was a transform through which Math had to flow. In that I did not have any training other than On The Job, the first 2 months were painful. Thereafter, my proficiency increased and tutoring became much more effective.

Engineering

In essence, Engineering == Problem Solving. Scenarios are too numerous to enumerate. My approach to teaching problem solving involves looking at contemporary methods, inside and out. Thereafter, I strongly advise use of the highest form of math possible with a swing toward graphs,

Analog Design

In this realm the dominant designs tend to be PLLs, voltage regulators, current sources, delay lines, Herein the emphasis has been on the essential elements along with technology challenges present in more advanced process nodes.

For example, I would tend to cluster PLL, Delay Line, voltage regulators and current mirrors and cite the hardest elements to produce with precision: resistors. High precision and invariant capacitors are another challenge.

Given these challenges, designs need to be tolerant of these uncertainties. That aside, there are other constraints in PLL designs that include input frequency limits, limits of VCO operating range. Testing of these tend to require skew of devices, ranging the supplies and methods to guard against uneven device skew.

Standard Cell Design with DFT for Undergraduate Level

In 1983 scan design was an IBM secret and undergraduate education with rare exception (RIT) did not include design and manufacture of chips. I wrote texts and developed some curriculum elements to propagate principals of Schematic Capture, logic synthesis, design for test and scan testing at the undergraduate programs for Penn State, Purdue, University of Illinois and University of Minnesota. (IBM VLSI Academic Program)

Notable attributes were 4.2 mm chips designed one semester and returned to campus, tested in 10 weeks time. This was an unprecedented use of former IBM secrets, a tight coordination with the fab to get chips done on time, along with a custom designed schematic capture program, designed to run with 256kb ram and an optional floppy disk drive which could hold 712kb,

MBIST

During a collaboration between IBM, Motorola and Apple, (in 1992) we jointly developed PowerPC. At the time, for the high end 64 bit unix processor chip there was a confluence of memory designs which had inconsistent access methods. I devised a single scan based test engine that would exhaustively test all memory elements without adding any delay in any functional data path.

In essence, this machine embodied a precursor to contemporary OCC designs that not only had a clock pulse counter for a million cycles but also governed shift enable in the same as functional clock speed. This was used to insert address, data and controls into datapath elements by "issuing a shift instruction" just prior to a functional clock for testing. This run for n cycle clock counter also afforded us cycle repeatable debug engine. (patent : Apparatus and method for testing a memory array)

In this effort I mentored a few folks who subsequently became master inventors.

Physical Design

Starting from Power4 processor I led the effort to assemble and integrate a single core 64 bit processor in 1998 from an original dual core application (Power4 Chip Integration – Hot Chips 12)

In this process I developed two additional team members to take on a role of integrators. This included training on pin and bus placement code I wrote in Perl.

Finite Element Solving

In the world of computing, a simple technique to calculate a time dependent function is called finite element analysis. In this method (

https://books.google.com/books?hl=en&lr=&id=JsCg-QWUT28C&oi=fnd&pg=PA1&dq=finite+element+analysis&ots=jRzuSMswjP&sig=apjKmnUM1c1BSnKTS_nl_eRNu10#v=onepage&q=finite%20element%20analysis&f=false)

In essence this allows a vector math operation to be represented quite easily and with almost no memory requirement. My first such programs were in the realm of pn junction device analysis where the research goal was to empirically derive turn on voltage by projecting the curve from "off state" forward while driving the curve of the linear region backward. Theoretical results predict the intersection of these curves to be turn on voltage. This research was conducted under the guidance of Henri Banerjee who authored a very concise test on Modern Physics while at Rochester Institute of Technology. Computer available was a mainframe with ~64kb memory, standard in of punch cards, standard out of paper.

Later, as an IBM employee, I had the task to support finite element programs which required "vector processors" to model and simulate semiconductor processes and devices with great accuracy. Author of these codes was Kevyn Salsburg a former colleague and subsequent team leader.
(http://prabook.com/web/person-view.html?profileId=226854)

Students of this course of study were process and device physicists within Fairchild, Motorola and other large companies.

K-12 Math and Science Outreach (National Engineering "Week")

Technically, this national program was and still is an annual effort to reach out to students and convince them that pursuit of studies in Math and Science will open doors for lifelong success. In that vein, there is a declared week which begins with Washington's birthday holiday in the us and consumes the week in school visits. To comply with the spirit of the program merely means you send some folks on campus that week and do whatever you think may help.

Given my passion on this, that approach was deemed to have too small a multiplier and too steep an entry point. While I could find countless people unhappy with the current stats on how few study math and science, most were reluctant to participate for one principal reason. They feared they would mistakenly do or say the wrong thing and correspondingly turn off the student they intended to encourage. In essence, the problem was too unbounded.

Thus, I devised a means by which volunteers could attend a lunch in which they could conduct an experiment intended to exercise problem solving, math and science elements and group thinking. In these training sessions a problem statement would be introduced for 5 minutes. Groups of 4 would form to solve the problem, They were allocated time to reach consensus which needed to be drawn on a sheet of paper, Upon consensus, they would begin solving/building for 20-25 minutes. Once time was up, demonstration of each solution was conducted by each team.

They then had 5 minutes for optimization. Final trials then determined a winning team but all students received a prize. The best part of the prize was the debriefing to show how each team used math/science.

The outcome was stunning. We went from former reluctance of volunteers to 600 per year reaching 25000 to 30000 students. The window of time grew from one week to a semester. Many students reported intention to add more courses of math and science as they felt validated in the tangible fact that they COULD.

Design for Test (University of Maryland, College Park)

Early in my IBM career I decided I wanted to pursue a Masters Degree in engineering. I thus enrolled in several courses at Maryland. During this era I was also helping teach at University of Illinois, Purdue and University of Minnesota. Each time I tried to take a course, I was asked to lecture on topics. I was also so active in my teaching schedule that I failed to have time to devote to taking classes. I did enjoy the survey classes I conducted at Maryland as well as the others I helped enable at the aforementioned Universities. At the time, there were no real available texts to address Design for Test at the level of Undergraduate while there was some material available for graduate programs.

And The List Goes On

To suggest an artificial limit of the aforementioned method of teaching implies that I have found a counterexample. Try as I might, this method seems to work in any setting with any student or any subject area expert. In essence, where there is a will, there is a way.

The underpinning of this method has it history in the age old methods of master/apprentice, teacher/student, parent/child. Key to understanding and application is there is no rocket science but rather respect and rote practice.

Comments Are Welcome

Any comments you have, suggestions for improvement or desire to further discuss can be directed to me @ edwardmseymour@gmail.com

Jack of All Trades

As a closing observation I make no claim in being a subject matter expert on any of the enclosed topics but I have an uncanny ability to teach. Further, in my travels, over time, I know it is time to move on when people with greater proficiency attain the confidence sufficient to teach.

Ok, So What Is Next, Future Aspirations

My current aspirations line up with offering a fresh approach to teaching with the intention of moving methods to ones offering higher gain or efficacy. In this next section I will enumerate more specific areas of focus.

K-12 Education for Math and Science

Circuit Design Techniques For Pure Tone Processors

Circuit Design For Maybe and Native Decimal Operations

Design For Test With Simple Comprehension

Fill In The Blank

Penny For Your Thoughts

Think Bigger

www.ingramcontent.com/pod-product-compliance
Lightning Source LLC
Chambersburg PA
CBHW041118180526
45172CB00001B/315